The Grey's Monkey

Ammed Williams

PUBLISH AMERICA

PublishAmerica
Baltimore

First printing

PublishAmerica has allowed this work to remain exactly as the author intended, verbatim, without editorial input.

Hardcover 978-1-4512-8605-2
Softcover 978-1-4512-8606-9
PUBLISHED BY PUBLISHAMERICA, LLLP
www.publishamerica.com
Baltimore

Printed in the United States of America

Table of Contents

Explaining the Anthropomorphisms

The first of the principals to be voiced in this writing is that all of the anthropomorphisms that have arisen on the planet have all come, first from fish, and then subsequently from frogs, toads and lizards. The second principal is that these species branch off into tree varieties and ground varieties. The third principal is that these tree varieties stem into gliders and climbers. And the fourth principal states that every being, once thought a fabrication can be accounted for with these basic facts in mind.

If we bear in mind the first principal, that all anthropomorphisms have come from fish, we may arrive to the body type of an angel by focusing on the predication that the feathers, do indeed come from scales, which a primitive primate—due to the third principal—that these reptiles branch off into gliders—would possess and adapt into a flying apparatus, if an ice age were to produce white skin, hair and plumage and brain capacity which should also produce the necessary behavioral characteristics of an angle. Such being's remains have been found in digs and we nevertheless have a record for them in our mythology. That of course implies, that all primates have originated first from either frogs or lizards or perhaps even toads, but much less certainly so, from toads, than the former.

There is a second part to these principals. Species revert to past traits when conditions become less favorable for their current phenotypes. This stands as the first principal second part. Secondly to these principals, within the second part, there are seasons to the planets evolutionary conditions. There are ice ages—or winters, springs—or lushness of those survivors of the ice age, summers, or downfalls of lushness due to the third part (as yet to follow), and Autumns or cooling periods on the globe, once life has been devastated by the summer season on the globe. Again, this will be called the second principal second part. The third principal, second part, states that during the winter season on the globe those that are smallest become dominant due to sheer benefits of a less demanding energy consumption quota, than do currently dominant animals, which sink in size progressively and then regain some size in the global spring. Along with this the planets summer is meant to weed out gluttons through famine, and the autumn is to prepare those that survived summer haste for the long winter of changes in mass. This stands as the second part, third principal

But this is not all, and there is yet a third part. The first principal is that through the act of breathe by those dominant summer animals the seasons of the ages are themselves interrelated with animal life. Oxygen intake combined with CO2 exhalation, further combined with use of plant resources by whatever dominant animal, which produces the oxygen no longer, leads to the cooling period and subsequently the ice age. This stands as the whole of the third part.

We now have the forth part. The only principal of the fourth part is that animals of whatever size, in an ecosystem, as they should approach the winter age, take on less mass than their pregnant mothers did during their own gestation. Reversal to this as they should move towards the peak of the spring age, their daughters will eat better when they are pregnant than they did, so gaining mass with successive generations. This inverse relationship between ice age approach and receding is the whole of the fourth part.

So, with these relationships and parts in mind, we may bear on the questions that folklore have asked of us, and make use of our new found approach to questioning origination of lesser known species. Take, for example, those species of giant humans found in digs. If we bear the principal in mind of the fourth part-humans, as they approached the global spring of their dominance, mentioned before in the second part, were able to take on more mass with each successive birthing. This ended up resulting in the display of massive characteristics attuned to giants. There was a back lash, quite speculatively in the summer which approached from over eating and excessive CO2 exhalation, which plants could no longer provide for, and thusly, the winter made these giants less grand in size, whist retaining their exaggerated features. We may now see how these principals lend themselves to the phenomenon in folklore, only evolutionarily so, pointing to trolls. But, perhaps we may also draw on the first part, the origination of frogs, to account for the phenotypes displayed by giants, after their summer and into their autumn, which may have crumbled a civilization, and produced what we have mentioned as the first principal (second part)—or reversion into ogre characteristics from former primitive DNA. The recovery from this display, may well have been ground dwelling trolls, whom, with exaggerated characteristics, are less large than ogres, and, of course, less large than giants are taken to be.

Imaginatively, as there is no branch of truly devoid science as of yet on the matter—we may investigate the origination of any of the varying displays of mans likeness. One of the harder things to fashion, without the old brand of fossil finding, is how to determine, without layers of sediment in the earth to guide you, what the epochs are for human life, that a child of tomorrow may see in a futuristic science book, the way we have no doubt seen in so far as the depiction of dinosaurs. Such phantasmagorical story boards, are nevertheless dormant in the human psyche, but are to be accurately drawn by terms of a dating system, whereby reptilian hominids, yetis and Grays walk the earth in one epoch, angles and flying devils—the anthropomorphic grandsons of primitive lizards, in another.

What are the implications of this tenet of anthropology? You might say mythology becomes better adapted to practical aspiration, except these beings, which have been accounted for to an extraordinary degree in writings of the past, are no longer here to serve as challengers to human life in the present. So we may look to those civilizations which were bound to crumble at some point in history and learn from their mistakes when we take on our own reversion as it has to come at some point in the course of our development. It is possible to suggest that by analyzing the fact that giants have primitive counterpart—ogres, they were once perhaps civilized, in order to account for the mechanistic de-domesticity, and thusly, human civilization ran counter. Since the phenotype of the giant has tried and failed at Turn and Nod on earth, that humans will go the way of the gnome in our eventual makeup, as this has been accounted for to some successful degree in England's phenomenon may be a possibility.

All beings of old, and those which fall out of the scope of any science which does not recognize these principals, are ascertainable through deliberation and thoughtfulness, as one may refer to principals of the first part when one must account for reptilian or amphibian humanoids, the first part in the case of tails, having been derived from lizard's subsequences of fish, and later the fourth part may account for the varying sizes of all the beings of the past and current folklore, while the third principal first part may account for all those of the winged variety on the point of flying capacity. The advent of hair is something we leave to be fairly implicit in life, but we may consider it a certainty that the mammalian condition is little more than its parts thrown together, as we know it to be today.

Hair can arise without live birth as is indicated by the platypus, and live birth has been displayed in reptiles, such as some specific breeds of snake, and as far as warm bloodedness is concerned, it may be free from live birth and hair together, as with modern birds. So hair, warm bloodedness, and live birth, are secondary to other phenomenology, as we may account for all of them by referring to a creatures benevolent ice age response, which may only arise if the animal had been progressing through natural evolutionary practice henceforth. Such is evident in the piecing of the cases to follow:

Starting from fish, at what point in their history did angels become markedly mammalian in nature. We must first consider that humans display the phenotype of having a use for their arms in dexterity. While it is the lizards which retain the scales necessary to lead up to feathers. Also, added to this, angels must have undergone ice age conditions to amount to their tell tale white appearance which, as in polar bears, yetis, and arctic foxes is used for not being seen unless they choose to be in snow environments. So, if we start from fish, progress to climbing lizard, arrive at gliding lizard, approach, which by all rational must be a certainty based on our apparent anatomy, the genetic makeup resembling that of a primate, then make our way, through the knowledge of their tails to flying devil, undergo an ice age, which can account for the feathers of the once scaled wingspan, and integrate a cross breeding of the human gene of the dexterous arm, we may find ourselves with an explanation for angles which is rather satisfactory. While this is by whatever means apparently useful, we must, when trying to ascertain that vision of a human epoch storyboard, whether some flying devils were to remain with their current phenotypes, or by some abominable hellish civilization collapse, revert more to their lizard inheritance and go the way of the dragon in a later display of genes. We must admit, that dragons seem to have some wit about them, characteristic of less than dinosaurs, and while not as much, a little more of anthropomorphism.

We have set out a brave explanation of angels, as well as devils, but we become lost as to our current understanding of the role of apes in the matter. For this, we must redefine apes. And also, we must redefine the standing apes have in our own personal history. The most primitive of primate species still exists in some form today. It is known as, by one variety, the slender Loris. And by all logic, it must have come from a tree dweller, with many of the same characteristics. The principals of the first part already state that it must have heritage as a frog, but this animal bears a resemblance to the tree frog like no other. It undoubtedly ends the debate as to primate inheritance. Loris's have backs that seem to be merged with their faces. It is as though they lack necks, and they hunt insects. Their faces are somewhat pointed. They are nocturnal. Their eyes are reddish black. No doubt they are mammalian, but one only needs an ice age to put an end to that characteristic dependent most probably, on the size of the reptilian primate in question, for it could not go underground in such a circumstance as having to dig more than topsoil would allow.

Loris's have four fingers, which is something they also share with frogs. They use sticky secretions, and they climb trees. Their climbing is identical to particularly the red eyed tree frog, which shares all of the above mentioned characteristics with the exception of both size and mammal character, which may both be accounted for by referring to both the third principal second part and the before accounted unimportance of marked mammalian tendencies.

Thusly, it is by observation of red eyed tree frogs phenotypes along with Loris's that we may see for ourselves that the linkage of frogs to primates is, even today, all too apparent. To further exhaust the debate, humans, during our development in the womb, undergo a resemblance to the common tad pole; which surges the idea that at one point, our species took a turn toward live birth, from their former egg laying as some particular marsupial frogs have been known to do to some degree.

These dealings will be extenuated in proceeding chapters.

Optics and Anthropomorphisms

The utmost employee of the spiritual factor and a revolutionary idea of brain hemispherical cosmetics is the use of mirror optics on a person's eyes. I have used myself as an both experimenter and case study to determine human types based on the active or passive quality of my or any persons eye nature. These, for practical purposes may be labeled Yin—mirror or Yang—fog. This chart illustrates the eight inherit natures of man and his hemispheric nature and preference:

If a person with a Yang Yin disposition were to go to a mirror, and were to gaze into his right eye, standing very close to the mirror, but not so close as to be unable to see the full right eye, and if one were to look into ones right eye, with ones right eye, and then only, ones right eye looking into itself, the eye would then change its nature, and become more yang, until the eyelashes blur and then take on the full appearance of a yang/yang personality. Optics and neurology tells us that when one see a shadow, or a mirror, one begins to operate ones consciousness in the alternate hemisphere. If one is using ones right hemisphere, and gazes at ones shadow, one begins to then use ones left hemisphere and vics-versa.

It is by doing this for only one eye, that one achieves the imbalance necessary to change ones appearance and the nature of one's eyes. What cannot be changed, although vision may be imparted to either hemisphere, is the use of the hemisphere as a preferential. These optic anthropomorphisms come most handy when one considers the before lain spiritual factor.

Celebrities are those whom are iconic of their sidedness. John Goodman, a yin/yin, nevertheless, uses his left eye predominantly. George Clooney, also a yin/yin, uses his right. There is a marked difference in psychology also impingent on other factors between the two, but nonetheless are intrinsic to the type. To change one's eyes changes ones opinion, not only of one's self, but of one's fellows, and tremendously alters ones viewpoint of those one had hardly considered worthy of befriending before. The worst scenario is coming back to a split yin/yang or yang/yin without knowing yin from yang. This is a matter of understanding the paradox of laterality that is seeded in the nature of our bodily consciousness. It is easier, far easier, to be born yang/yang with a left eye preferential, or the Hyde figure than to live with one's self by such a methodology.

How to Account for the Combinational Mythologies

Hypothetically, there was once a period in Greece when humans did not eat omnivorously, or even agriculturally, but may have became specialized predatorily. If we but examine the case of modern flamingos, their inheritance is widely dispersed to the genetic inheritance of their prey, or brine shrimp, yet they resemble and differ from other birds primarily on the point of their vey localized diet. If one were to put a flamingo in captivity, it would still retain its coloration by way of thought experiment. Thusly, the pink coloration of a flamingo shares both of two parts.

The sixth part, first principal, is that by digestion, through the digestions small villi, new proteins are taken in through sub sequentially imparted nutritional incorporation by any of the cell's nucleus. The nucleus has been said to be both responsible for the actions of reproduction as well as protein synthesis, which in turn may go to constitute the nucleic acids to follow by way of the digestion process. When a being is widely set apart from another's origination, the implications over generations are minor—such as the vastness existing between shrimp and flamingos, pandas and bamboo, but the creature nevertheless incorporates that which its specialized diet will permit—or the full scope of colorations on a flamingo where extremities are appropriate, or the genuinely benevolent vibrations that are lent to pandas by their incorporate bamboo diet. The second principal, sixth part states that through specialization, if the object of one's specialized diet has a common ancestor with the eater, body diversities may be resultant. The whole of the sixth part may be coined digestional-traditional-somatic inheritance or the passing down of food specialization over time, to amount to a species origination.

Since both man and horse stem from fish, while horses had then gone a separate way, in geologic time it would make little difference when one must account for the display and record of the Centaur—which by proposal, was a descendant of a tribe of man to eat his horses, and underwent what we may call phenotypic-specialist evolution.

Phenotypic-specialist-evolution may account for other periods in history, when an animal such as a wild goat, may provide the area of specialty, and provide explanation for the evidence of the horned hominid skulls we have located in a dig.

Those individuals with the predatory skill to hunt the more formidable of wild life may result their biology according to such extravagance, but, on the point of being individuals, result in the monsters that heroes were said to face. It is imaginable under these precepts to account for all the classifications in the Greek annals of once thought inexplicable, anthropomorphisms.

It is also, through conviction of this, that one may assert the tendencies of dogs, to resemble and understand humans, is founded on their wolf inheritance, in a time where much DNA may have been shared to those canines by way of phenotypic-specialist-evolution. One may hold the same true, but to a more primitive degree in the cats, whereas they are more distantly removed and more primitively displayed, regarded by some token to be soothing and well as showing mal-intent.

And so, through the diet of the specialist, genes may, in an ongoing population, be activated as phenotypes that were never inherited by way of direct ancestry.

This makes a great impact on the progress of finding explanations for all of the diversity of human life, and the anthropomorphisms.

How to realize the First Human Epoch

In the very beginnings of human history, when the first human types were being introduced, as has already been stated in the first principal first part, the beings which were to have arisen were to come out of frogs. We go from climbing frog then proceed to reptilian primate. Most likely, reptilian primates descend from the trees and see their way to being bipedal reptilian primates, as to evade the dinosaurs whom were capable of flight at the time. What is then produced is the reptilian hominid. These creatures, more than likely, existed before the Grays and before any mammalian primate. They most likely had a civilization, which collapsed in the global winter, due to the second principal, second part, and were then evolved, due to the relationship between global winter and mammalian traits, productive of yetis, which, today are, in warm regions, now bigfoot. In swamp regions 'creature of black lagoon.' Those of the reptilian hominids that withstood the global winter became mammalian, and thusly human.

After the ice age, which most probably, crumbled the reptilian civilization, the reptilian primates, which were still tree climbers, became mammalian, and exist today as nocturnal Loris's. The Loris's diversified and produced apes, as well as varieties which retained the characteristics of the Loris as a biped, and, towards the spring, and the mass change of the second part, third principal, wind up with a civilization and an impressive brain size, as indicated by fossil evidence. These Grays, develop technology, and eventually space technology, and, although being unable to escape their home solar system, do return to their home planet earth, having undergone the time dilation one may account for by understanding the principal of the twin's paradox, an, already established, branch of physics. One may also account for the pyramids, and pharos with this evasion of the past by traveling into the future, although primitive humans and Grays, even civilized ones, lived about the same time. Their offspring, in what moderns have called 'mixed relationships' are the elves.

The modern phenomenon of abduction is an effort to continue the fate of astronauts of the Grays, by hybrid breeding through artificial methods of egg and semen extraction. The product of this, in the modern world, are so called 'alien, human hybrids' and eventually, when Gray DNA has undergone many generations, fathering and mothering, you have parties of Grays whom appear totally human, except for their elf like beauty.

Those of the reptilian civilization, whom were space travelers, undergo time dilation to wind up from their present, to their future, our past, and abduct Grays for the above mentioned purpose.

To return to the point in question—the epoch—by way of the sixth part, first principal—these Loris, possibly while still being reptilians, most likely were, even back then, insect eaters. As I have personally seen footage of a Loris's sharing a praying mantis, I think we ought to draw on the bamboo eaters of our time, in order that we may speculate on the nature of these creatures…insect like eyes.

As well as this, there is a species, of space going Gray, which has reverted to its reptilian primate heritage in the collapse of the Gray civilization and domesticity. Behind this fact, those Grays who were not astronauts were wild in the Gray collapse of the global winter, due to the second part, third principal became smaller, due to the second part first principal, reverted, and became the modern chupacabra of Mexico. Gray space going parties are, nonetheless, dependent on earth's resource, subject to reversion, and regardless of which, sometimes are confused with the chupacabra in name.

To coin a biophysics principal for this phenomenon of UFOs, one may say, biological inertial frames, whereby, perhaps even, Reptilian hominids, began, drew from other species—Grays—and then underwent speeds for distances at times necessary to continue their race drawing from their own direct descendents, or man. In other frank terms, UFOs are the contraptions of our own ancestors who first came from fish.

Reptilian Hominids and Domesticity

Since—due to the extent of their technology, it is conceivable that reptilian hominids, being our cousins, had a civilization, it may be pertinent to look to the domesticity of one other animal—the pig—to pinpoint our history's advent of mammalian characteristics, and particularly our pink skin and its type. Since pigs require a fostering that is not nomadic in its nature, and has permanent shelter as a requisite, along with plant agriculture in its terms, which is fixedly detrimental in location, pigs are the start of an area fixity that domestsication lends to the evolution of pigs, as being previously bores, now softer of skin, and—through digestional traditional somatic inheritance marks the transformation of reptilian hominids to humans of today. The fetal stage of humans, as having resembled so closely that of a pig bears on some of these suggestions as to the actualization of the genes which—once genetically synthesized—become phenotypic of the species natal development.

The Second Human Epoch

The second human epoch is far less alien than the first, far less primal; actually more spiritual, and on the side of religiousness. If we continue, here, our discussion of the flying humanoids—angels and flying devils, and also take it in mind, to consider bipedal lizard-primate humanoids, or devils, we may come into the picture of this eschatological time on earth. All devils, since they have tails, by the first part, are descendants of gliding lizards, then gliding primates, then flying bipedal primates, then flying bipedal humanoid primates, then officially flying devils.

We must take it into consideration that biblical accounts have little to do with the evolutionary standpoint, as the only way to account for a feathered winged human—or angle—is by a flying devil which has undergone a global winter, retained feathers from scales then cross bred with a human over time to gain arms. This, in effect, renders angels descendents of nighttime visitors in name of the succubus and incubus, or devil like creatures whom would have sex with people, throughout history, in sleep, rather real and biological rather than occult. One would do well not to do too much piecing of the bible by eventually fully angelic incubi, that would be a mistake, as the trinity does hold much importance as a matter of symbolic depictive metaphorical prose for mans soul journey. It is a mistake to read too far into the biological account of the Christ figure.

To account for man goats, we must consider that sixth part, and understand that it would only take human predatory goat specialization to produce these.

This is pretty much conclusive of this period, though I would like to add that I am not nearly as involved with the details as I have been with the first human epoch. May this be regarded as scientific and not deconstructive of what should be read with connotation only in mind in the first—first place. I would like to point out that angels are not apparitions in their truest form, and may be killed just as devils may be. The fact of the matter is that they mark a separation between frogs and lizards, rather than good and evil. Let that be the sounding point.

Angels and devils have been covered, bipedal devils have possibly interbred or eaten humans to a degree, that doesn't tell much of a success story when they are no longer capable of flight. As there is current phenomenon related to dog children, it is possible that by eating an animal, one makes interbreeding more possible, though the dog children are completely non functional in the cases we understand to whatever degree, as would be a bipedal devil, in some regards. The practical applications of digestion inheritance, followed by sexual reproduction by the same, are of course, completely unethical by modern practicality and, in one's ones children, manifest non-functionality. I imagine research of the dog children may prove, perhaps, that such is in as much as accurate to practical degree.

Though we may consider ourselves religious, we may now have a scientific ornament with which to understand our idea of biological opposition, and that is the tail. While lizards are dinural, frogs are nocturnal, and while one may consider it an opposition, it is more of a co-existence between the Yang aspect and the Yin aspect.

The touchiness of this chapter, cannot be forgiven, and should be tackled rather by purely biological accounting for individual species—however—I have remained impartial to the understanding of heaven and hell (especially early on), and so do not have the sufficient understanding of their members.

This should, much more gracefully be done by a naturalist which has some understanding of images of the associated artistic period with the principals herein, in mind.

The Third Human Epoch

Dragons, in the medieval period, perhaps, and surprisingly, may be the descendants of devils whom which were to only eat humans, and so gain the body type of arms, as flying devils are inherently without them, and the personification of dragons may be genetically resultant in said diet. They of course, due to their size were to be quite successful, and due to the fourth part, arms would have been an easy evolutionary step when considering the sizes they maintain in legends over the course of their eating. The vestigial nature of their arms is also indicative of a digestion inheritance by way of its afterthought. Green witches are those solitary—or wild/non-domestic humans typical of the period which, in turn, and due to the second part first principal, have reverted to their reptilian hominid ancestry. Angles, due to the fourth part, thereby undergo a change in mass over time, due to the global winter prior, and in the spring of the globe, manifest once more as ferries. Having been winged creatures, it may have made sense to specialize in flying insects and incidentally undergo, upon a change in mass, the phenotypic display that would permit their smaller size flying ability derived from their dietary inheritance of flying insect wings. Giants manifest by way of the fourth part, approaching the summer before the third human epoch, and then recede progressively into trolls dwarfs and finally gnomes as Global year's progress. Ogres are reversions of Giants after the global autumn and may display some reptilian makeup.

Elves are the offspring of Gray/Human interbred ancestry. The ears in elves are such that they account for the highly developed cerebral cortex, while maintaining the human need for an auditory organ residing in the mammalian brain.

Demons are elves-as indicated by their pointed ears—which have reverted to reptilian DNA—as indicated by their green skin. The supernatural nature of a counterpart goblin, is in as well as due to its reverted ancestry, which is in itself, perhaps demon reverted elf displays newly undergone due to its grotesque attributes. A gremlin is a reverted elf in its makeup, which has undergone, as a result of the fourth part, a shrinking during a famine of global winter, from an earlier, larger, demon display of size.

Witch Exert

To speak of witches in its most traditional of senses, we typically register a haired green female. To get an understanding of why the picture most always results in a female we may have to draw on some common knowledge about the reptilian nature of a witches makeup co-factoring with her sexuality. Perhaps we may begin to see in our makeup of today, the revert nature of the genitalia—as operable in sex changes, and the persistence of the nipple across the sexes to find that with our medieval reversion back to traits of the first epoch—came an adoption of that trait's sexuality under the strain of the civilization collapse. Hair was retained from the mammalian development, while reproduction may have become the responsibility of a sole feminine parent. This is displayed today to some degree in frogs, and would make an improvement on the understanding of the retention of a male nipple—should these evolutionary strides be repeated—as a vestigial element. As alternate to this, we may decide that the sex acts among witches were perhaps more hermaphroditic, thus maintaining genetic diversity. Whichever the case, it stirs the imagination to see life at its most creative.

While not every species becomes mammalian—thus gaining lactation—not every mammal becomes domestic, and thusly, not every mammal reverts back to its prior standing, or even has one of which to speak. And so—should a sexuality, typical of a specie's reptilian, in this case, reverted self, demand the traits of a female, across the board, it would follow—regardless of the traits mammalian nature—that the species, once reverted—displays the morphology as such, should the trait be feminine. Humans, if theory holds, are unique in this manner on the point of the nipple, having come from the singularity of this specie's sex, reptilian in nature, second to, and consequent of, the traits mammalian introduction. The nipple is therefore a visage of our witch ancestry and therefore reptilian hominid descent in its modern realization on the human male, having endured (and come out of) our specie's reversion. For no other species should this be true, lest there be some unseen factor with which to add to the curiosity.

Anthropomorphic Animal Deviancy

It is possible to suggest, and supported by phenomenon, that animals of anthropomorphic nature, such as bears and dolphins, may be, in the case of bears, descendants of the frog family to have gotten closer to civilized than appearances tell, in their current phenotypic displays. They may be descendants of native shamans, which have switched to the diet of their ecosystem, as an omnivore, and thusly displayed the phenotypes of the native wildlife across the board, thus producing animal qualities. It would then follow that singing is inherently, a pragmatic thing to do around a bear which has forgotten its manners.

Dolphins on the other hand, may be the descendants of shipwrecked humans to have become so specialized in fish, that they withstand a change in body type, and over time, come to retain lingual abilities, but, lose their human likeness nearly entirely; the modern species a descendant to mermen.

The Wealth of Grazer Phenotypic Displays as Product of Predatory Specialization throughout Human Eras

While in one instance, trolls may have reverted once from Giants, their heredity is nonetheless rooted in humans, whom consumed much wooly mammoth in the instance of the global winter. It should therefore follow that the phenotypic display of tusks in trolls, and possibly ogres, stems from that digestional-traditional-somatic inheritance.

In instance of the second human epoch, one way to account for the redness of many of the satanic figures, is by proposing that they feasted, as a co-causal factor, on the blood of Grazers, and became predatory of such. This produces, perhaps, the red tint of the skin, as well as the DNA of the horned head, which varies with the animal of specialty, and is possibly morphed between specialties and sexual inheritance. Animals such as the Bison, Impala, and Gerenuk may serve this basis, insofar as art typical of the period concerns itself. Another proposal as to the color of a satanic figures skin may stand as a sun tanned feature, while yet another may be due to the artificial connection between the underworld, and its environment, having a biological root in clay, and thusly being caked on to one such satanic figure from its nightly habit of underground shelter. All inasmuch are possible, but as of yet not entirely ultimate in instance.

Curiously, the tailed quality of modern grazers signifies that they have come from the lizard family—and hence may—due to anthropomorphic animal deviancy-be lizard family descendants of the anthropomorphic beings of the second human epoch to have stemmed from bipedal devils. This however is unclear. If we take into consideration the horizontal pupils of a satanic being of the second epoch, they are tell tale signs of their digestive traditional somatic specialization, an evolutionary arms race trait, now turned on its originator—the grazers. These pupils are used, through evolutionary time, for the purposes of a broader range of vision non-peripherally. It therefore follows that grazers, though probable descendents of an anthropomorphic being, are not necessarily the products of anthropomorphic animal deviancy.

However, the practical assertion for why the gene of the horn would become phenotypic is a weaponry against any rival, either of one's own species or otherwise. There are also contentiously for the purposes of sexual selection among parties to the species and those that would produce a successful partnership throughout anthropomorphisms.

The hove of the satanic beings of the second human epoch can be indicative of the very gradual range of displays produced in anthropomorphisms, which could, in all actuality, lead to a modern animal deviant or an entire classification.

The Fourth Human Epoch—The Modern Age

Never, since the pyramids, have humans been able to achieve such feats of technology. Then it was due to Gray intervention. One might say that this equal yet advanced intelligence is responsible for the feats we are achieving through the nurturing and tailoring of the human mind on the part of the Grays for the benefit of both Grays and Humans alike. We are now, speculatively, though not without merit, in cooperation with the Grays and building an empire that both may one day inherit in subsequent DNA incorporation. The laws of mans God still apply, as well as Karma and the basic idea of taboo and totem, sin and atonement. One has trouble understanding the grand schema, and one incorporates his or her own missteps in what one will later on have to figure out for oneself, perhaps vicariously. Everything is more advanced than ever. And we are on our way to a rather confusing but progressively nurturing, though lonely era. Because of the fact that intellectually, we are not alone, we tend to feel nonetheless isolated by the fact that we never witness separation. Only a multitude can witness another. When the two are one, they seem alone. A lonely era, though progressive, is what we have on our hands. There is no submission, no overcoming, only incorporation.

Television as Cross Human

Since elves are those beings— which, without the aid of biological inertial frames— integrated with human's intelligence their own natural ability— as well as that brought on by bi party nurture— their harmony, it was done in successive instance with mans plight. Currently we receive the aids of the past success stories; spearhead of man's ancestry: aided by genetic makeup of a cousin, and then technologically reintroduced into the earthly populous— revolutionizing daily the way we behave through mimetic info— which can do harm as well as helping. Life lessons are learned, through the technological era, by way of television, being but a man made illusion, which has responsiveness to inner suggestion, but, no recourse to the byproducts of what was—by nature—left veiled, speculatively— as intended. Join in an elf dance and never be heard from again. We may be a culture, wherein even a genetics major— with a history minor— would have a hard time talking his way out of the above accusation towards the spirit of television with conviction, especially if he had cable. This'd be due— of course— to the paradox between wanting to defend television—though its use— . . .and then knowing the spirit of television well enough to see it presented within these pages. We become lost in the light of the television; so much are we lost that we do not reach for our lamps. It is consuming. It has become who we are. It could not be possible, were it not for some understanding of that telepathic side of paradise's fences.

Tails

The tail, originally developed for the purposes of propelling and water navigation, becomes—as animals embark onto land, used for the purposes of fat storage. Thusly it should only follow that tails become an emblem of the sinful. Greed in terms of the intake that only becomes reprimanded to a useless appendage, lust in the admiration of another's ability to possess and consume affectively, envy in the admiration of foes acquiring of resource over one's own, wrath made in the display of the trashing of the tail as to indicate the excitements of fury and the boasting of past successes, pride in one's own display of fitness as typified by the tail, and sloth in the fact that the tail may always be resorted as one need not strive in times of harsh facing, and of course the fat gluttonous tail speaks for itself. But it is the fact that the tail has been made, primarily for show that renders it a function of the weapons of the psyche. This is not to declare that animals with tails are sinful— but it is meant to demonstrate the inclinations of our race to have viewed the anthropomorphisms with tails as having been.

Digestive Somatic Inheritance from the Perspective of the Right Wing Sociological

If we examine the link between DNA protein synthesis, by way of digestion, and the need of encompassing a single individual to derive full benefit, auxiliary to the need to benefit from multiple diversified individuals, the need arises from a sociological standpoint to allow the predominance of a typical processing plant, which may incorporate multiple processed biomass individual, into a single unit which may be purchased, such as a can or jar of vegetables or fruit paste, shredded leaves or otherwise of multiple plants and the like. This would produce upon consumption, a large library of genetic information, whereupon disclosed to newly produced cells, and over seven years—alter the DNA of an individual tremendously and healthfully.

The purpose of this section is not to spearhead lawful enforcement, but to forward personal recommendation by the healthful living among us socially. Such is as well as in the status of healthful countrymen, but not devoid of international application. Therefore, all whom would understand need benefit from the applications of diet to digestive inheritance.

Optics and Anthropomorphisms Extended

Keeping in mind that the universe is dual in its aspects, one may derive that in a single person, each possessing two eyes, and these eyes wired to both hemispheres, as the nature of the eye itself is either operatively wired to the left or right, two eyes times two possible hemispheres equals four, four types times two, or the number of residing instances of consciousness—equaling eight—then compounding the four types making use of both hemispheres simultaneously— equaling twelve—then compounding eight mirror function switches for each eye preference type, along with eight in the opposite direction of switching for each eyed preference, then compounding twelve more for the shadow function on the four types with eight eyed preferences—equaling forty—then eight more acting on both active hemispheres by way of mirror for four, and shadow for the other of the four types, arriving at forty eight pairs, plus one which has no consciousness at all, amounting to ninety seven types of eyes . If a person possesses a yin/yin type they may be in use of—in terms of consciousness—their right eye, connect to the left hemisphere by way of having been a yin/yin, and having last looked in a mirror when they were previously operating with the left eye—connected to the left hemisphere—also by way of the all too stated yin/yin type, and displaying an introversion in the eye to have looked it should produce in the eye to have looked an extroversion while changing the consciousness to the other eye. It is believable that the mirror puts a person in the extroverted, while a shadow puts the opposing consciousness operative in the introverted.

Thusly, in a person born with, or artificially induced to produce—a yang/yang function, if they were to just pop into existence displaying introversion in the left eye, without their consciousness residing there and extroversion in their right eye, with consciousness residing there, and the consciousness wired to the right hemisphere, as with both hemispheres on account of the yang/yang disposition, they would then be inclined upon a meeting with a person of another disposition to, be the basis of the very basic most idea of the interaction, display controlled extroverted masculinity. If, however, they were to look into a shadow, they would thereby display, by use of the left eye display the introversion on the part of both eyes, with control over the left hemisphere. Where and when a person uses both his hemispheres simultaneously is a mystery I—quite uncharacteristically—hope to never solve.

Fusion and Innovation

The laws of new genetic acquisition are governed from the joint influences of fusion in the case of chromosomes, whereby two—previously dormant—recessive traits, become activated by way of correlated coincidental correspondence, and in the cases of DNA synthesis by way of ingestion followed by digestion into the cells of a single genetically gauged individual, along and in cooperation with the innovation of a new organism, to have inherited all the previously lain synthesized store-housing. Therefore it follows that new traits are made apparent, only through fusion of the sexual, which is dormant and previously recessive, by means of the digestive side of the fuse. Mutation therefore is more in so resultant from these two actions in relation to each other.

Food Intake Correlated with Cell Division in Weight Gain

If we hold it true that cells have a lifespan, and divide approximately 50 times before each dies, than those cells to divide will also have a lifespan and divide fifty times before they die as well. Under the current model, we supply these cells with the basic nutrients necessary to live. But what does this statement mean? If a cell need only take in basic nutriment to exist, and if we then examine the bottom of the food chain can we there find something that cells necessitate that are gotten from anything but other life forms? Plants take in nutriment from the soil which is only posits of dead matter. The sun and water and a third unspecified component seem to be the basis for achieving cellular homeostasis. What if, in order to divide, a cell needed to acquire some outside DNA in order to produce successful new cells in the inner ecosystem of the body, and thus lent this role to the digestion of food and introduction of new amino acids into the blood stream, thereby reaching the cell, arriving at the nucleus undergoing a synthesis, and then allowing for a unicellular division of a type particular in its DNA to its tissue function.

Thusly, one may not obtain nutriment lest that nutriment provide genetic diversity.

The reproduction of a single celled unit, in a bodily ecosystem, is dependent on, in its asexuality, the integration by way of blood nutriment, the necessary amino acids, in the necessary quantities, to result into fit new cells with the old cell as a genetic basis, and having been allowed a genetic advantage in the ongoing evolution of its nonsexual reproduction. Sexual reproduction draws form ingestion too, but on the basis of the sex cell, all genetic information is then potentiated in the conceptive process.

Cells, requiring little actual material supplementation which may not otherwise have been provided by water or optimal function of the organism through interexchange, should only be depended on for the sake of adding biomass, if only the organism can ingest a component which is environmentally lacking inwardly—to allow for the success of one's cell's eventual division. Thus you have a prognosis for health matters which are not systematized to a balance or pyramid of every available supplementation

It should follow that all the food chain may be linked in such a way as this—whereby each organism provides its predator with the building blocks of its own prey, in a fashion to lend diversity in both directions of the ecosystem their fair share of genetic variability in order to replace existing cells, through their inner evolutionary arms race and so continue their existence.

If a person was not a vegan, and wanted to lose weight in a particular area, it would be sensible to find food products in the meat section that did not share much in common in the way of its anatomic likeness to the area one desired to deprive of cellular division. It would be too tempting for the body to have ready adaptive amino acids used in protein synthesis by the time they reached that part of the body, and therefore one should not give ones inner network a chance to take such a shortcut. If one makes it harder for cell division to occur, than those deprived areas will operate at or approaching normal, until they simply must be fed genetic information necessary to continue function of the tissue. This of course is the closest thing to health that the body need provide. The most convincing and possibly most ridiculous of arguments for the importance of these factors must be a remark of a man eating a dog tail— "When I eat the meat, I am happy."

Communicative Ability in Species

As chemotaxis is the capacity to direct oneself towards a given smell, so as to apprehend ones prey and thereupon eat it, communication therefore evolved along these sensory modems in own ability to control the output stimuli of one's own physical apparatus. This could be called, in the case of humans—self incriminating audio-taxis, arising from that first and utmost vulnerability to being eating by giving oneself away by way of sound waves—to a potential predator. It therefore follows, as more species are able to control their output stimuli, audio-taxis itself becomes interchangeable by way of conception of new stimuli, and reception of perceived stimuli, based on one's history, and allowing predator and prey to change places, and fluctuate roles, interchangeably as submissive and aggressive self controlled displayers of what were once things to be hidden for the sake of survival. When a species gains control over what it had been concealing from its predators, and reveals it to its peers or encounters, it may thereby affect that party with the innate chance to gain some as immaterial benefit, thought it is actually the mark of the chase—be it sexual or predatory, and therefore—inherently: the bi-pronged approach to spiritual and biological incorporation via partners of interspecies and species alike.

This may as well be seen in what may be called photo-taxis and its self incrimination of the deep sea fish by way of light, and in those which are able to adapt their scales or skins in the cases of squid—chromatophores, this becomes self incrimination of one's position, as well—thus teasing and taunting another to become more receptive to ones aggressive behavior, when necessary—be it sexual or otherwise predatory, though less typical. It therefore follows that females, if the less aggressive, males if more aggressive, should be the less ordained in ways of entrancement on the part of their physical apparatus and its capacity for control, females if more aggressive, males if less, in the opposing natural didacticism. These natural didacticisms of a species modal-taxis self incrimination culture may teeter with generations, and each generation is responsible, in part, for the mistakes of the last, especially in terms of the good of the species as a unit, deeply rooted in the mind of the organism and seeded in what may be found to have a protein correlation in DNA.

Hair Acquisition

It is suggestible, though latent with speculation, that the evolutionary advent of hair came about by means of, be it the ingestion and nucleic synthesis of the bristled insect life, thus transforming a more primitive pangolin into a haired ant eater, over generations of ant consumption, as already progressive in a pangolins display of neck bristles, along with the inspection of ant bristles via magnification. This may in as well, also be the case where the consumer targets things such as bristled horseflies, or perhaps plant life such as the kiwi fruit or the coconut. This may found the genetic information, only to be used in case of cold climate. But where cold is atypical some phenotypic likeness may occur whereas the coconut crab displays both the coconuts coloration and bristles. This is to indicate the idea that the haired coconut engenders hair in the crab which eats it.

Coloration

It is the ecology of a region that will specify the availability of different diets, and so allow a typical body type its origin of particularities. It is not so much due to random mutation, one should think, that a squirrel regional to one area should display red bristly hair, while another, with a different available diet should produce mutations of grey hair. Much more logical, one should think, is the factor that the availability of their food should determine what phenotypes are made available by genetics. Further bolstering the idea, it so happens that when prey is consumed with some sort of physical design, that designs may be seen as apparent in the predator's phenotypic makeup. Camouflage becomes self evident, where a species need not accidentally produce mutants with genetic favorability to a region, but, instead, eat commoners in said region that are sustained by a food chain that is in common with all member's coloration.

Lightning Source UK Ltd.
Milton Keynes UK
14 February 2011

167502UK00002B/348/P

9 781451 286069